Scattered groups of rocks are seamlessly
emerging from a soft, sandy, desert-like
landscape that fills up the field of view,
or rather, the picture.

      With familiar shape, but unknown
      composition, they are, at the same time,
      a landscape,
      a sample,
      an obstacle,
      a value.

Here, on what seems to be the surface of the Moon, the rock is a reminder of our shared geological belonging and the capacity of world-making,

no matter whether it fits into a hand or takes the form of a planet.

The ground is mostly flat, covered with
dark sand and shaped by soft dunes.

        The landscape feels settled
        and calm, until there is movement.

Controlled,
slow rolling of the wheels,
almost imperceptible, not stirring up the soil.

The digging, drilling, and scraping,
on the other hand, triggers a warning:

    a storm of blinding dust that
    makes everything and everyone stop.

The light is coming from a strange place,
lying low, filling up this dark space
with more shadows than brightness.

It seems that, here,
shadows play a special role:

      the contrasts help the machines
      orient themselves,

      they determine their choreography,
      indicating the material conditions
      of their environment.

      Despite the intensity
      of the glow,
      the temperature remains glacial.

On one side, the ambiguity of the light
coming from either the Sun or the laboratory,
on the other, certainty:

in both cases, it feels like a theatre
stage, strangely devoid of predetermined roles,
open for interpretation.

The darkness,
real or simulated,   decontextualizes
this space from
its surroundings,
obscuring its temporalities
and histories too.

Once a distant subject of legends and myths,
it is now an immediate object of

scientific investigation,
financial fantasies,
and colonial imaginaries.

The soft landscape is occupied by machinery, roaming around, and claiming the territory.

Navigated via remote-control systems of networks, signals, and ambitions, these rovers are finding their way between obstacles—whether topographical or technological in nature.

Robotic hands are scratching the surface, searching for gold.

Moving,

    falling,

        getting up, digging,

    people are helping.

This carefully designed and executed performance
is not put on for applause.

It is information, shinier than the Moon itself,
that the scientists,
        economists,
        politicians,
        and astronauts are looking for.

        That they collect, study, and translate
        into data and possibilities of discovery,
                profit,
                growth,
                and glory.

This lunar surface, *real* or staged,
operates both as a map and a territory.

Distances between the machine and the rock,

the rock and the human,

the human and the Moon,

the Moon and the Earth,

the Earth and its desires,

the ultimate connectedness of planets,
objects, subjects, imaginaries, pasts,
and futures, is contained in one tiny
sample of this performance.

        The naming is essential—earthly techniques
        of appropriation are put
        into orbit to claim the cosmic.

THE MOON AND THE SCREEN

MARIJA MARIĆ

The Moon is back. This time in higher resolution and a different media format. Photographs of fearless astronauts bouncing off the lunar surface in their carefully designed space suits, searching for an adequate place to plant their national flags of technological superiority and colonizing capacity, have been replaced by computer-generated images and videos of heroic technologies — rockets, rovers, robotic arms — all automated, remotely controlled products of a rapidly growing space mining industry. Communicating what has commonly been described as the New Space Age, referring to the recent renewed interest in space exploration, now led mainly by private companies and guided by a more "pragmatic" vision (that of mining for rare metals and minerals) — the space mining media has occupied the digital domain as one long advertisement, with no beginning or end, neither geographical nor cultural specificity, no history, only future.

This shift from photography towards advertising as a narrative domain in which space mining is being proffered as a scientific, but also financial, political, and social project, should not be seen as a mere historical consequence of the evolution of media technologies. Instead, it testifies to a more structural change: the shift from nation-states and international organizations to private companies as main stakeholders in space exploration (and appropriation).

Furthermore, this shift from a singular powerful *image* (e.g. the photograph and television broadcast of the 1969 Apollo 11 lunar landing) towards a *genre* of space mining video trailers, renderings, and animations could also be seen as a genuinely infrastructural turn; infrastructural not only in terms of its focus on the technological products and services marketed in these advertisements, but also in terms of the genre itself as a form of media infrastructure, able to — through its invisible, universalist framework of familiar narratives, short, anonymous content, *lingua franca* English and online dissemination — legitimize and naturalize our understanding of the Moon as a new resource frontier.

THE MOON ON THE SCREEN

The rendering of the Moon as a new resource frontier, however, is not new. Starting from the seventeenth century's rationalization and first scientific explorations of space — technological progress and the close-up view it enabled have radically changed the ways of seeing the Earth's natural satellite. In other words, the Anthropogenic framing of the Moon started with the process of its mediation and its translation from Moon-in-the-sky to Moon-on-the-screen. The invention of the telescope in the seventeenth century could be seen as the first moment of *screening* — looking at the Moon beyond the optics of the naked eye, with technology as a mediator. Finding its place in front of the telescopic lenses, the Moon has effectively become *studied*, turning the cislunar space into a planetary laboratory. The telescope allowed astronomers such as Galileo Galilei and Thomas Harriot to produce the first lunar maps, which in turn inspired further advancement in the domain of optics engineering, resulting in an ever sharper and more magnified image.[1]

The invention of the daguerreotype in the nineteenth century by French painter and printmaker Jacques-Mandé Daguerre allowed for the first technical reproduction of the Moon, thereby shrinking its spatial, temporal, and cultural distance from the Earth. The first lunar photographs, such as John William Draper's 1840 daguerreotype of the full Moon, which successfully documented its terrain, craters, and rock formations, inspired scientific analysis more than artistic appreciation at the time.[2] Enabling a closer, more detailed look at the Moon, early photography essentially turned it into an object of study, positioning technology as a necessary mediator of this relationship. As such, photography did not only free the hand by delegating its artistic function to the eye looking through a lens but has also permanently blurred the vision of our plain sight of the Moon.

The mediation of the Moon, however, took a significant turn during the Cold War period, initially with the 1957 launch of the first artificial satellite, Sputnik 1, into an orbit around Earth, and then with Luna 2 in 1959 — the first spacecraft to successfully make a lunar landing — both delivered as part of the Soviet space program. Beyond the technological achievement of the two missions, Luna 2 — designed to release a 650 kilometers-wide bright orange sodium vapor cloud upon its impact on the surface of the Moon — was a particularly significant project, as it represented the first human object successfully recorded from Earth. Soon after, Luna 3, a spacecraft equipped specifically to collect images of the far side of the Moon, captured 17 photographs, thus marking the first in situ documentation of lunar landscapes. But the Soviet space program

had a further achievement: it launched the Moon as a matter of global news, which, in the context of the Cold War, started a geopolitical (and media) competition to become known as the "space race."

However, it was not until NASA's Apollo 11 mission, which brought humans to the Moon, that popular media became an essential part of space exploration. Equipped with hand cameras based on a technology devised by the Scottish electrical engineer John Logie Baird—who also developed the world's first live and color television—Neil Armstrong broadcasted his first, choreographed steps on the lunar surface, thus marking the Moon's true screen debut. With an estimated audience of 528 million people, the 1969 lunar landing could be considered the first truly global media experience. In his analysis of the NASA space program through the lens of the Apollo 11 spacesuit fabric design, Nicholas de Monchaux effectively demonstrated the interconnectedness of television and the military-industrial complex in the USA. Arguing that "the entire effort to go to the moon should be rightly understood as an elaborate apparatus for the production of a single television image," he indicated the power of a singular media event in shaping global politics.[3] Comparing organizational structures of NASA with those of CBS, the news network in charge of the live broadcast of the lunar landing, de Monchaux showed how the years of preparation for this 31-hour long television event, including the making of the "bank" of related footage, correspondents, simulations, and background stories used in a non-linear, non-scripted way, also set the ground for the media world as we know it today.[4]

The termination of the Apollo program in 1975 was followed by decades of relative silence, shaping a period characterized by a transition from documentation to imagination, from fantasies of landing to those of settling. At the forefront of this shift was Gerald K. O'Neill, a professor of physics at Princeton University, whose visions of human settlements, produced in collaboration with visual artists Don Davis and Rick Guidice, and published in his 1976 book titled *High Frontier: Human Colonies in Space*, came to serve as the visual iconography of the post-Apollo era. The foundations of this project were set during the 1975 Summer Study organized by O'Neill in collaboration with NASA and Stanford University, with Davis, Guidice, as well as Peter Glaeser, an architecture critic and curator at MoMA.[5] O'Neill's space settlement project was grounded in the emerging political and economic questions of its day—energy crisis and resource scarcity on Earth—and shaped by the techno-utopian design and engineering of the time.[6] The fantastic images of space habitats, such as the iconic *Cylinder*, managed to capture the imagination of space colonization, lending the hostile and distant outer-space environments a homely and familiar interiority.[7]

While O'Neill's paintings represented space colonies as idealized landscapes on Earth, "with repeated allusions to the Californian coast, the island of Bermuda, Italian villages, and the South of France—all familiar vacation spots for privileged Europeans and Americans," as Felicity D. Scott observed, Glaeser's approach to mediating NASA's colonization project was characterized by his refusal to use architectural visualizations, focusing instead on the formats of system-based protocols and diagrams.[8] However, his understanding of architecture as a "mediating mechanism" that transcends the form of the built structure and instead appears as an organizational tool, as she further explains, did not imply the critique of the seductive imaginaries of colonization, but rather "a refusal of such aesthetic supplements while simultaneously attempting to render the apparatus legible."[9]

Beyond the 1970s' turn towards techno-utopian visions of space settlements and bureaucratic aesthetics of architectural organizational power on Earth and beyond, post-WWII space exploration was a genuinely popular media experience, in which photography and television played a key role. The construction of spacecraft, space media technologies, and imaginaries of space colonization went hand-in-hand, operating inseparably. The documentary character of photography and live television broadcasts played an important role not only in attesting to the truthfulness of the important events on the Moon or proving the technological achievements of the two main competing nations; their media character also served as a tool in the campaign of their political regimes. In this context, photography and television became mediators of interplanetary distance, effectively translating the unknown into the known, fiction into reality,

This shift, now commonly referred to as the New Space Age, could be seen as a consequence of changes in the legal framework of different countries, most notably the USA, Luxembourg, Japan, and the United Arab Emirates, which accommodate the private tech and mining companies operating from their territories.[10] The liberalization of space mining laws in these countries, often understood as a controversial interpretation of the "Moon Agreement" and the "Outer Space Treaty," essentially allowed private companies to legally conduct research and mining operations on the Moon.[11] The New Space Age thus effectively represented a shift from nation-states and international institutions to private companies as main actors in the process of the exploration, or rather, exploitation of the Moon.[12]

With the shift of stakeholders, the media formats in which space activities and imaginaries are communicated have also changed: photography, television, and the fantastic images of outer space colonies have now given space to a more "pragmatic" format—that of advertising. Appearing mostly

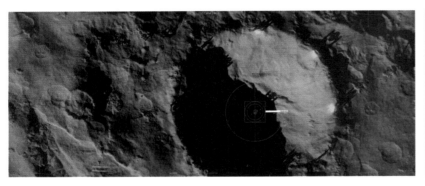 

and generally setting the ground for the claiming and appropriation of the mediated sites and worlds.

THE MOON AS THE SCREEN

The decades after the end of the Apollo missions saw little to no activity in terms of lunar research, except for NASA's Clementine missions of the 1990s, and the later involvement of other countries in space exploration, such as China or India. During the past decade, however, interest in lunar exploration has returned, now under the premises of space mining—the exploitation of resources on the Moon and on asteroids.

in the format of short videos and animations, the New Space Age media narratives have largely been organized around demonstrations of the cutting-edge technologies necessary for space mining or its associated activities like transport, material processing, energy infrastructure, to name just a few. Characterized by high-resolution computer-generated imagery and animation or shots of rovers at play in the decontextualized environments of staged lunar landscapes, space mining advertisements feature not humans, but rather technology as their protagonist, or technological and financial ecosystems suited to the development of such technology, accentuating rupture, innovation, and discovery.

The representational realism of technology is well coupled with the dreams and fictions of profit and growth, cast in the constructed emptiness of the lunar landscape. Immersive and repetitive, formulated as a staged showcasing rather than a storytelling project, the current media representation of the New Space Age could be seen as one long commercial.

Observing how all of the smart city projects appear not as fully functional cities, but rather as a "version," a "demo," or a "prototype" of urban models, Orit Halpern, Robert Mitchell, and Bernard Dionysius Geoghegan frame the "demo" as "a form of temporal management that through its practices and discourse evacuates the historical and contextual specificity of individual catastrophes and evades ever having to access or represent the impact of (...) infrastructures, because no project is ever 'finished.'"[13] Similarly, the space mining industry's short, digital-born, "demo" videos could be seen as triggers of technological and financial imaginaries of futures enabled by tangible technologies and services provided by the industry itself.

Moon, down on Earth. Emerging around the world as default testing grounds and infrastructural features of many space research institutions and private companies, lunar laboratories have become more than mere spaces for scientific experimentation and the testing of mining technologies. Instead, they have begun to occupy the role of media studios, where the imagery of these technologies "at work" are being produced—a visual element necessary for the operation of the speculative industry of space mining. As such, lunar laboratories and their artificial landscapes made of polystyrene rocks, basaltic sand, black curtains, and theatre lights could be seen as genuinely political spaces, "system[s] of literary inscription" in which collaborations between scientific research, private capital, myths of nation-making, and media apparatuses allow for the construction of industrial dreams of endless growth.[14]

Created in the shift between sites of the space mining industry's visual discourse production, the New Space Age advertisements are also a consequence of broader outsourcing of the production

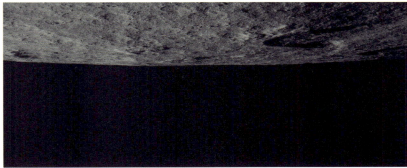

The move from documentation towards demonstration, characteristic of the New Space Age media, was a consequence of not only shifting focus but also shifting locus of lunar investigation. While the more difficult to achieve, in-situ activities such as the 1969 landing on the Moon and subsequent lunar missions generated attention through large media events, the quotidian, earthly testing of space mining technologies requires not only a different genre of representation but also different sites for the production of its media imagery. In this context of simulation, it is above all computer screens, software interfaces, and lunar laboratories that have appeared as new locations, digital and physical proxies of the

and management of media images to specialized external media agencies. The last several decades have seen not only the rise of in-house media and PR departments in large institutions and organizations but also of specialized media agencies, dedicated solely to the production and curation of their clients' media images. This transfer of media production to specialized agencies has provided media strategists with a lot of power in defining entire spheres of narratives, helping to universalize the products of these media strategies—in this case, space mining media genre—as a formulaic, repetitive structure that is itself industrially produced.[15] In other words, the commodification of the infrastructures, processes, and knowledge useful for the production of space mining media, went along with the commodification of space itself.[16]

If the post-war logic of singular events was focused on the making of the one memorable image or phrase, the current visual logic of the space mining industry is not based on memorability, but rather on the continuity of the repetitive visuals, all blurring into one single overwhelming environment in which information gives space to the atmosphere — an atmosphere of a legitimate and imminent technological future of mining the Moon. The infrastructural power of genre over the spectacular power of image thus plays into the invisible process of legitimization, and above all, normalization of the space mining project, enabled through this digital media surface. While the Moon had been on the screen before, the New Space Age has turned the Moon into the screen itself.

background for the staging of the technological products and services offered by the space companies.

Understanding the economy itself as a "tissue of fictions," Mark Fisher observes how "capitalist realism posits capitalism as a system that is free from the sentimental delusions and the comforting mythologies that governed past societies."[17] In reality, the opposite is the case. Similarly to other venture capitalists, space mining start-ups and companies must turn to more than mere scientific data, instead working with storytelling techniques, carefully crafted narratives, and all manner of fiction. In her book, *Friction: An Ethnography of Global Connection*, anthropologist Anna Tsing introduces the concept of an "economy of appearances," observing how "in speculative enterprises, profit must be imagined before it can be extracted."[18] Similarly, the economic performance of the space mining industry heavily depends on its dramatic performance — a stage on which science, finance, media, and theatre work hand-in-hand to build imaginaries of futures beyond Earth enabled by advanced technologies and private capital.

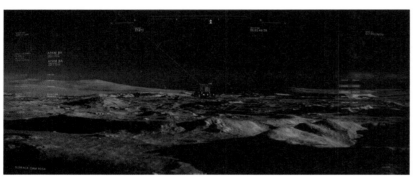
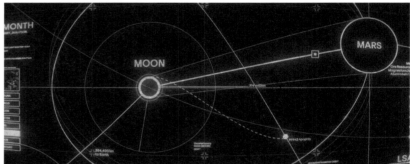

ATTENTION ECONOMY AND
MATERIAL FICTION

But what is behind the glossy, rendered surface of the space mining industry's media? Which kinds of fictions — narrative, financial, and political — are at the core of the New Space Age realism? Appearing as a response to different kinds of crises, such as that of resource scarcity, the space mining project never challenges the premises of the capitalist economy and the social, political, and environmental inequalities it produces, proposing instead alternative modalities for the continuation of the status quo. Here, the narratives of crises are productively employed as a

In other words, speculative economies like the one the space mining industry is based on, heavily depend upon attention, which appears as a valuable commodity in the context of global media and information-centered environment.[19] As such, space mining industry has a lot in common with the project of crypto-currencies, also predominantly shaped by a white, male, "tech-bro" culture and grounded in promises of "democratization" and radically different prospects for the future; the project whose economies are equally built on the notions of extraction and "digital metallism, that relies on the semiotics of metallic money, with its language of mining and rigs."[20] Media performance, organized around continuous rythms of booms and busts, represents an essential part of its financial logic.

BLURRING THE VISION

In his book *The Philosophy of Photography*, Vilém Flusser argued that the image cannot be separated from its technical apparatus, suggesting that the content of photography is also determined by the camera used to produce it.[21] Similarly, in her research of the digital language, media theorist Katherine N. Hayles called for a new focus on media-specific analysis, "a mode of critical attention which recognizes that all the texts are instantiated and that the nature of the medium in which they are instantiated matters."[22] To start unpacking the renewed interest in space mining, we need to understand its mediascapes—narratives, images, screens, dreams—and the politics that underlie them, especially in the context of unreachable sites of interest, such is the Moon, which force our dependency on knowledge of the media itself.[23] We need to look at the apparatuses and sites of their production—media agencies, software interfaces, lunar laboratories—in order to understand their actual content: earthly, material ambitions. If lunar laboratories and media studios are the sites where the rendering of the Moon as a new resource frontier is being produced, the question arises: what are the laboratories, or rather apparatuses, that could undo these visions?

In the context of late-stage capitalist production, and with speculative projects such as space mining, the media does not follow the project, but rather precedes it, serving as an active construction site where universes and universals are being assembled. The advertisements of the space mining industry could, therefore, be seen as sites where our relationships with the Other, or rather, with the *outside*, are being produced, where the constructed emptiness of the Moon intertwines with its constructed resourcefulness, and where its framing as a new extraplanetary mine is being naturalized. Their media-specificity as digital objects, produced by professional media companies, always presented in English with a repetitive aesthetic and thus able to circulate through online spaces without limits, becomes an essential part of the space mining project itself. Following on from this line of argument, it can be argued that it is not only the logic and modes of distribution which may be seen as inseparable from the distributed content. It is also the very genre as a recognizable form of communicating this content, which can be regarded as its invisible infrastructure. In other words, the media genre of the New Space Age, by constructing our ways of seeing, also constructs our ways of doing.

To challenge the dominant narratives of the space mining industry, we thus need to challenge the optical apparatus it imposes: to go beyond the "expert" view

and the close-ups and zoom-ins which precede the project of claiming and appropriation, thus rendering the Moon as an empty desert one can extract from. We must challenge the sharpness of this vision, and instead adopt the blurriness of looking at what falls beyond the optics of the Anthropocene; we must learn to see double, allowing for more than just one Moon to enter the picture again.

ENDNOTES

1. Galileo Galilei's *Sidereus Nuncius*, published in 1610 and translated as *Starry Messenger*, was the first published astronomical treatise featuring the drawings of the Moon and star constellations, inspiring scientists like Johannes Kepler to develop improvements in the telescopic lenses. See in Max Caspar, *Kepler*, trans. Doris Hellman (New York: Dover Publications, 1993).
2. John William Draper, *Moon*, 1840, The Met, London.
3. Nicholas de Monchaux, *Spacesuit. Fashioning Apollo* (Cambridge, Massachusets and London, England: The MIT Press, 2011), 147.
4. De Monchaux, *Spacesuit*, 257.
5. Fred Scharmen, *Space Settlements* (New York: Columbia University Press, 2019), 18.
6. Gerald K. O'Neill, *The High Frontier: Human Colonies in Space* (New York: William Morrow and Company, 1976).
7. O'Neill's interstellar Cylinder was a habitat-type of structure composed out of two cylinders, each 20 miles long and 4 miles wide, organized around three land areas, three "windows" and three mirrors, able to automatically open and close to create day-night cycles. See in Gerald K. O'Neill, *The High Frontier*.
8. Felicity D. Scott, "Earthlike," *Grey Room*, no. 65 (2016): 6-35, 7-8.
9. Scott, "Earthlike," 30.
10. The Outer Space Treaty, or *Treaty on Principles Governing the Activities of States in the Exploration and Use of Outer Space, including the Moon and Other Celestial Bodies*, which was drafted in 1966 by the Legal Subcommittee and adopted by the General Assembly of the UN in 1967, does not allow for national appropriations of the land on the Moon or other celestial bodies but leaves room for ambiguous interpretation regarding resource exploitation beyond Earth.
11. In 2020, *Artemis Accords—Principles for Cooperation in the Civil Exploration and the Use of the Moon, Mars, Comets, and Asteroids for Peaceful Purpose*, were signed between the US and several other countries in the world, putting a new emphasis on the questions of space commerce and the use and exploitation of space resources.
12. See also Rory Rowan, "The U.S. Geological Survey and Geo-Politics of Space Resources" (Presentation, Geography Seminar Series, Trinity College Dublin, November 4, 2020), https://www.youtube.com?watch?v=5e5dmAnOdVw&t=128s&ab_channel=TCDGeography.
13. Orit Halpern, Robert Mitchell, and Bernard Dionysius Geoghegan, "The Smartness Mandate: Notes toward a Critique," *Grey Room*, no. 68 (2017): 106-29, 116.
14. Bruno Latour, *Laboratory Life. The Construction of Scientific Facts* (Princeton, New Jersey: Princeton University Press, 1986), 105.
15. On the role of real estate branding industries in the mediation, design, and globalization of large-scale urban projects, see Marija Marić, "Real Estate Fiction. Branding Industries and the Construction of Global Urban Imaginaries" (Doctoral Dissertation, Zürich, ETH Zurich, 2020).
16. For a discussion of the way the rise of "image banks" affected the commodification of the image itself, see Estelle Blaschke, *Banking on Images: The Bettmann Archive and Corbis* (Spector Books, 2016).
17. Mark Fisher, "Foreword," in *Economic Science Fictions*, ed. William Davies (London: Goldsmiths Press, 2018), XI-XV.
18. Anna Tsing, *Friction: An Ethnography of Global Connection* (Princeton and Oxford: Princeton University Press, 2005).
19. Claudio Celis Bueno, *The Attention Economy: Labour, Time and Power in Cognitive Capitalism* (Rowman & Littlefield, 2017).
20. Nigel Dodd, "The Social Life of Bitcoin," *Theory, Culture & Society, Technologies of Relational Finance*, 35, no. 3 (2017): 35-56, 42.
21. Vilém Flusser, *Towards a Philosophy of Photography* (London: Reaktion Books, 2000), 21.
22. Katherine N. Hayles, "Print Is Flat, Code Is Deep: The Importance of Media-Specific Analysis," *Poetics Today* 25, no. 1 (2004): 67-90, 67.
23. On the concept of 'mediascapes' as "an interconnected repertoire of print, celluloid, electronic screens and billboards" see Arjun Appadurai, "Disjuncture and Difference in the Global Cultural Economy," *Public Culture* 2, no. 2 (1990): 1-24, 9.

Visuals taken from Luxembourg Space Agency video titled "Space Resources: Driving the Future of Space Exploration," originally published on December 16, 2020.

COSMIC MARKET
Armin Linke

The following series of photographs represent extracts from research conducted by Armin Linke with Francelle Cane and Marija Marić, developed within the framework of the exhibition of the Luxembourg Pavilion at the 2023 Venice Architecture Biennale. It gathers scenes of visits to sites, conversations, and archival documents encountered during the shooting of the film *Cosmic Market* (2023). All photographs were taken in Luxembourg and Italy during the filming period from November 2022 to February 2023.

    Collected between laboratories, space agencies, observatories, and space law offices, these shots bear witness to the different references, logistical components, and other internal communication devices that constitute the support of research around the Moon and the exploitation of its natural resources. They are a first step towards understanding the backstage of the space mining project, its visions, practices, histories, and futures, as well as the conditions in which its communities of scientists, economists, lawyers, and critics operate. Although individuals are largely absent from the shots, they are never far from the camera's lens, and their actions are always at the heart of the investigation.

1   Luxembourg Space Agency, Luxembourg, Luxembourg, 2022
2   University of Luxembourg, Faculty of Law, Economics and Finance, SES Chair in Satellite Communications
     and Media Law, United Nations Treaties and Principles on Outer Space, Luxembourg, Luxembourg, 2023

Article I *Free exploration, province*

The exploration and use of outer space, including the Moon and other celestial bodies, shall be carried out for the benefit and in the interests of all countries, irrespective of their degree of economic or scientific development, and shall be the province of all mankind.

Outer space, including the Moon and other celestial bodies, shall be free for exploration and use by all States without discrimination of any kind, on a basis of equality and in accordance with international law, and there shall be free access to all areas of celestial bodies.

There shall be freedom of scientific investigation in outer space, including the Moon and other celestial bodies, and States shall facilitate and encourage international cooperation in such investigation.

Article II *Non-Approp*

Outer space, including the Moon and other celestial bodies, is not subject to national appropriation by claim of sovereignty, by means of use or occupation, or by any other means.

Article III *IL, Peace*

States Parties to the Treaty shall carry on activities in the exploration and use of outer space, including the Moon and other celestial bodies, in accordance with international law, including the Charter of the United Nations, in the interest of maintaining international peace and security and promoting international cooperation and understanding.

Article IV *Non-militar.*

States Parties to the Treaty undertake not to place in orbit around the Earth any objects carrying nuclear weapons or any other kinds of weapons of mass destruction, install such weapons on celestial bodies, or station such weapons in outer space in any other manner.

The Moon and other celestial bodies shall be used by all States Parties to the Treaty exclusively for peaceful purposes. The establishment of military bases, installations and fortifications, the testing of any type of weapons and the conduct of military manoeuvres on celestial bodies shall be forbidden. The use of military personnel for scientific research or for any other peaceful purposes shall not be prohibited. The use of any equipment or facility necessary for peaceful exploration of the Moon and other celestial bodies shall also not be prohibited.

Article V *Astronauts*

States Parties to the Treaty shall regard astronauts as envoys of mankind in outer space and shall render to them all possible assistance in the event of accident, distress, or emergency landing on the territory of another State Party or on the high seas. When astronauts make such a landing, they shall be safely and promptly returned to the State of registry of their space vehicle.

In carrying on activities in outer space and on celestial bodies, the astronauts of one State Party shall render all possible assistance to the astronauts of other States Parties.

States Parties to the Treaty shall immediately inform the other States Parties to the Treaty or the Secretary-General of the United Nations of any phenomena they discover in outer space, including the Moon and other celestial bodies, which could constitute a danger to the life or health of astronauts.

Article VI *Responsibility*

ANY States Parties to the Treaty shall bear international responsibility for national activities in outer space, including the Moon and other celestial bodies, whether such activities are carried on by governmental agencies or by non-governmental entities, and for assuring that national activities are carried out in conformity with the provisions set forth in the present Treaty. The activities of non-governmental entities in outer space, including the Moon and other celestial bodies, shall require authorization and continuing supervision by the appropriate State Party to the Treaty. When activities are carried on in outer space, including the Moon and other celestial bodies, by an international organization, responsibility for compliance with this Treaty shall be borne both by the international organization and by the States Parties to the Treaty participating in such organization.

Article VII *Launch-states, liability*

Each State Party to the Treaty that launches or procures the launching of an object into outer space, including the Moon and other celestial bodies, and each State Party from whose territory or facility an object is launched, is internationally liable for damage to another State Party to the Treaty or to its natural or juridical persons by such object or its component parts on the Earth, in air space or in outer space, including the Moon and other celestial bodies.

Article VIII *Jurisdiction*

A LAUNCH-STATE A State Party to the Treaty on whose registry an object launched into outer space is carried shall retain jurisdiction and control over such object, and over any personnel thereof, while in outer space or on a celestial body. Ownership of objects launched into outer space, including objects landed or constructed on a celestial body, and of their component parts, is not affected by their presence in outer space or on a celestial body or by their return to the Earth. Such objects or component parts found beyond the limits of the State Party to the Treaty on whose registry they are carried shall be returned to that State Party, which shall, upon request, furnish identifying data prior to their return.

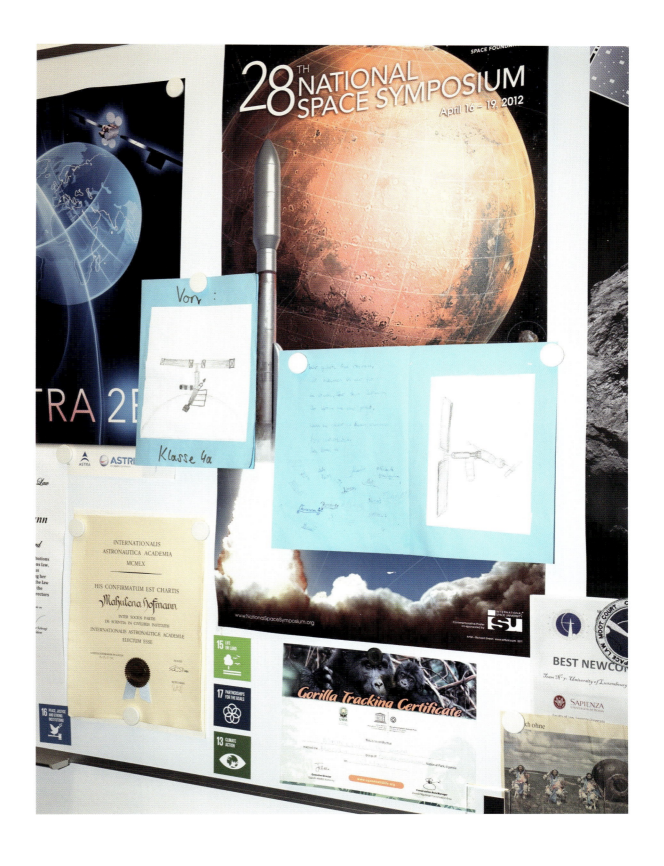

5 SnT University of Luxembourg, LunaLab, lunar analogue facility, Luxembourg, Luxembourg, 2022
6 SnT University of Luxembourg, LunaLab, lunar analogue facility, Luxembourg, Luxembourg, 2022
7 SnT University of Luxembourg, LunaLab, lunar analogue facility, Luxembourg, Luxembourg, 2023

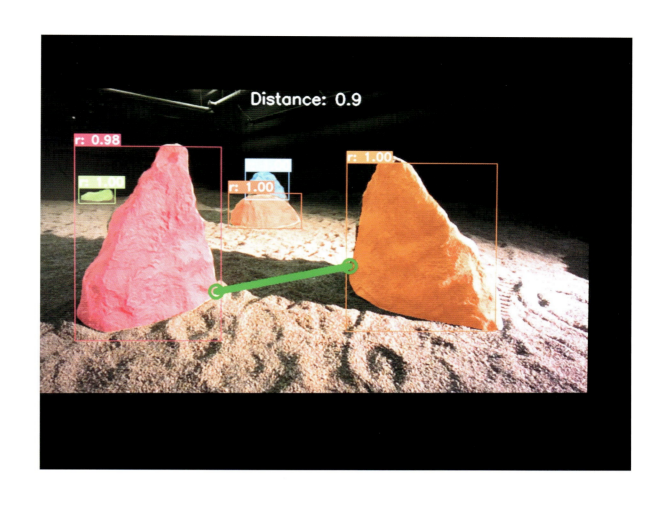

8 SnT University of Luxembourg, LunaLab, lunar analogue facility, Luxembourg, Luxembourg, 2023
9 SnT University of Luxembourg, LunaLab, rover training, obstacle recognition, Luxembourg, Luxembourg, 2023

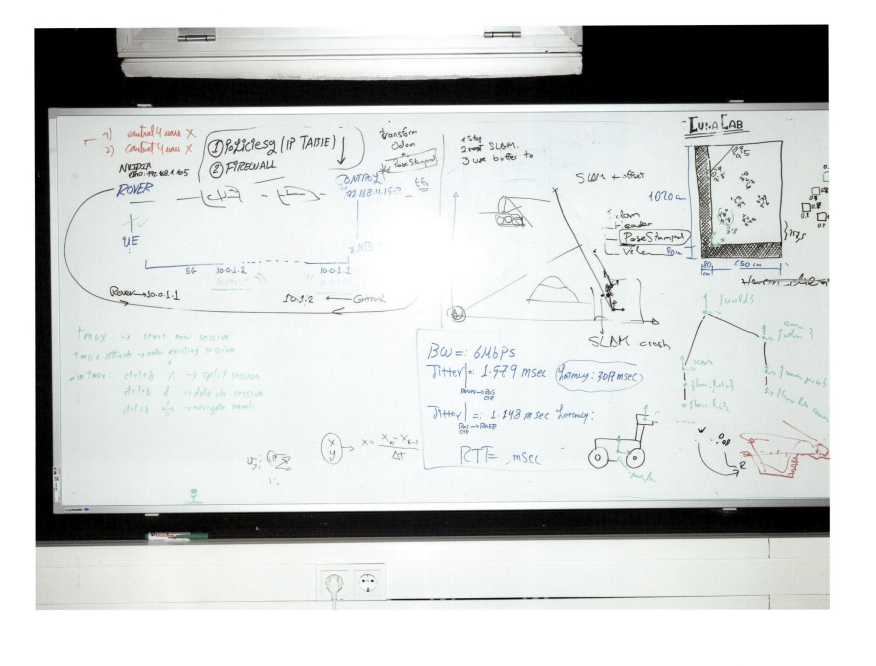

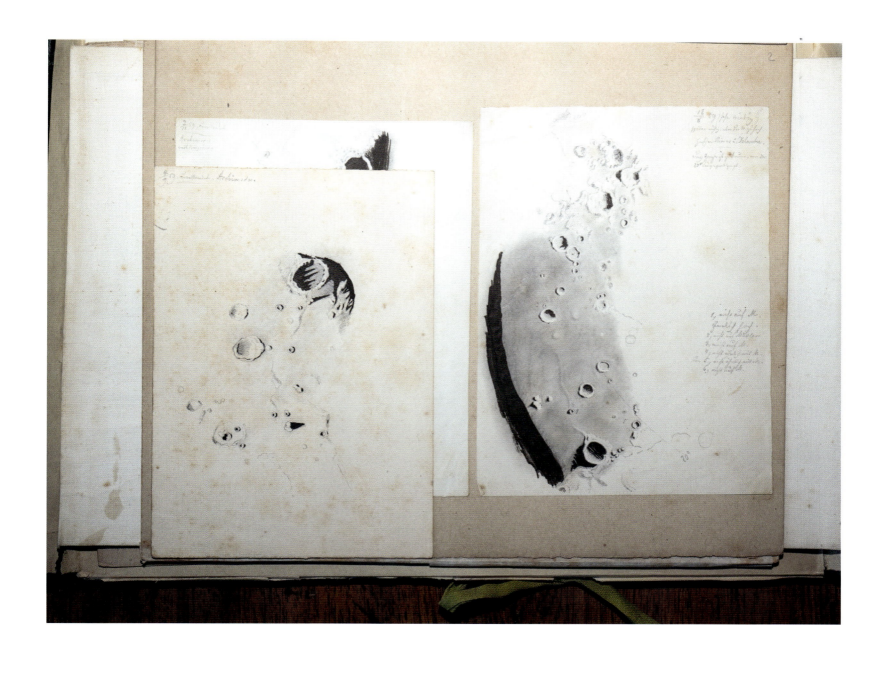

10 SnT University of Luxembourg, LunaLab, Luxembourg, Luxembourg, 2022
11 INAF (National Institute for Astrophysics), Astronomical Observatory of Arcetri, Wilhelm Tempel, drawings of the Moon's landscape, litographs, from the Historical Archives of Astrophysical Observatory of Arcetri, Florence, Italy, 2023

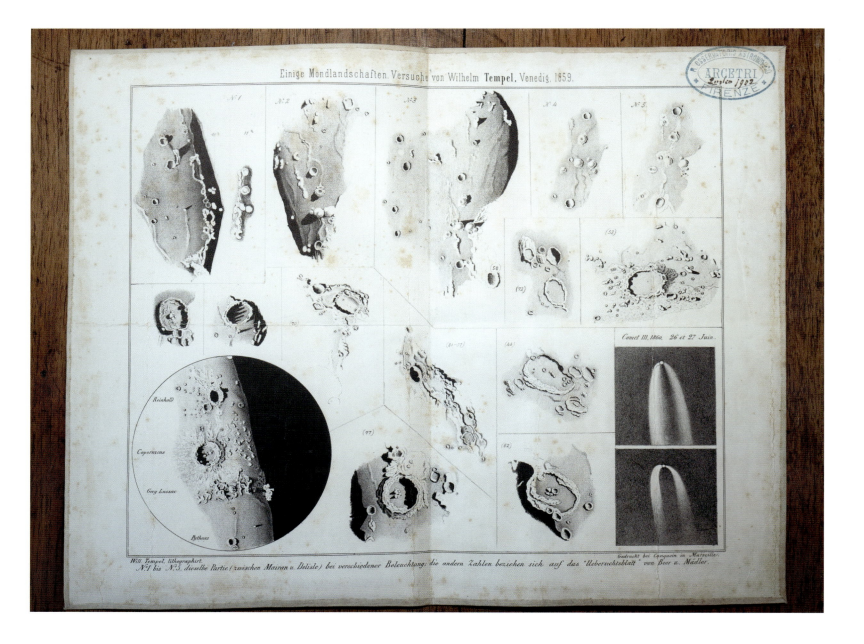

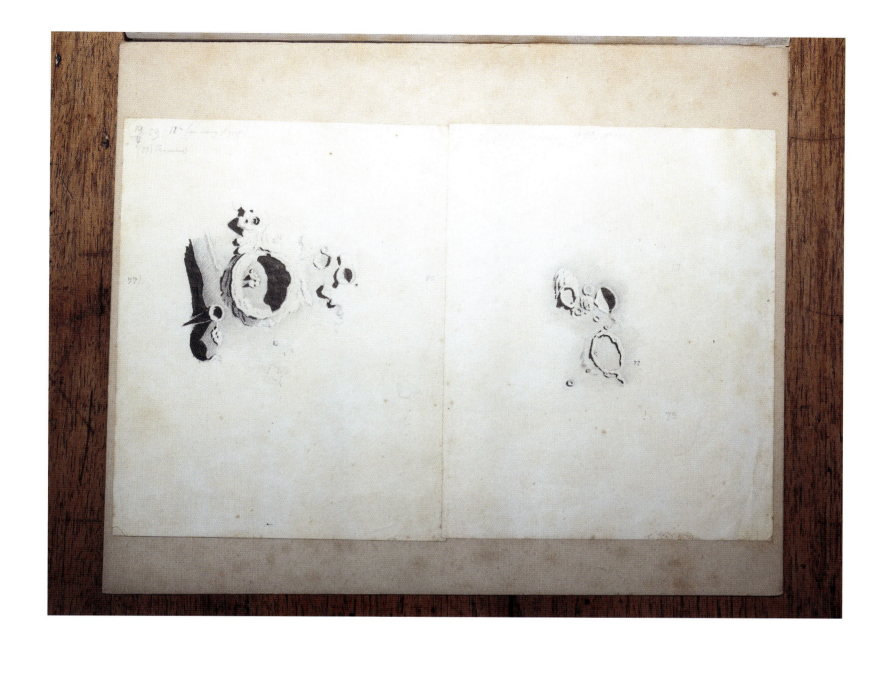

13   INAF (National Institute for Astrophysics), Astronomical Observatory of Arcetri, Wilhelm Tempel, drawings of the Moon's landscape, lithograph, from the Historical Archives of Astrophysical Observatory of Arcetri, Florence, Italy, 2023

14   INAF (National Institute for Astrophysics), Astronomical Observatory of Arcetri, Wilhelm Tempel, drawings of the Moon's landscape, lithograph, from the Historical Archives of Astrophysical Observatory of Arcetri, Florence, Italy, 2023

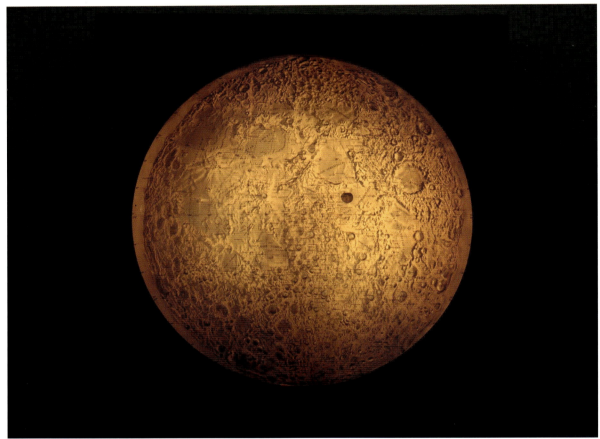

15   INAF (National Institute for Astrophysics), Arcetri Astrophysical Observatory, raised-relief map of the Moon, Florence, Italy, 2023
16   INAF (National Institute for Astrophysics), Arcetri Astrophysical Observatory, raised-relief map of the Moon, Florence, Italy, 2023
17   Francesco Esposito, INAF (National Institute for Astrophysics), Astronomical Observatory of Capodimonte, PowerPoint presentation on the potential of lunar resources, Naples, Italy, 2023

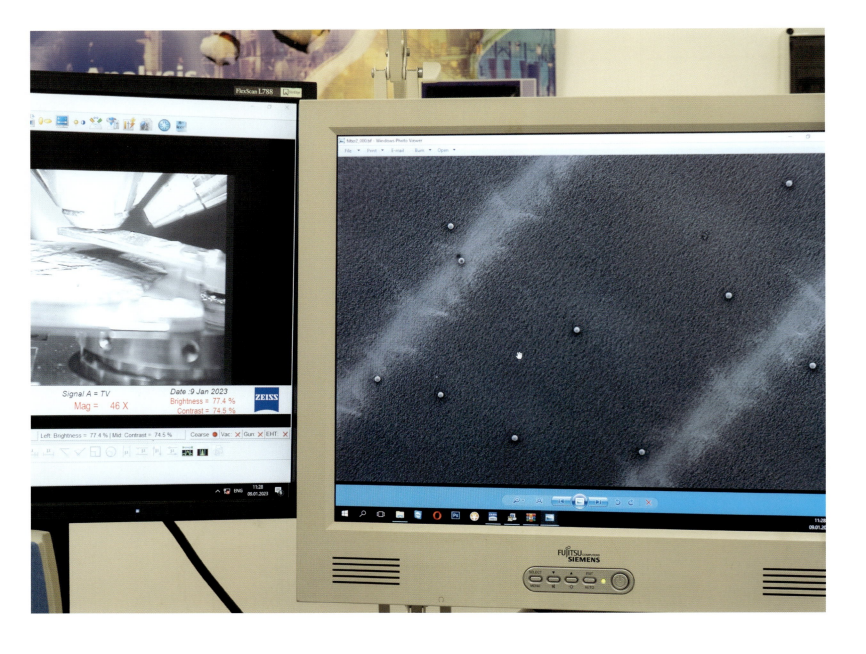

18 INAF (National Institute for Astrophysics), Astronomical Observatory of Capodimonte, development of optical dust particle counter for the Moon and Mars, microscopic observations of space dust, Naples, Italy, 2023
19 Maana Electric, developing miniature automated solar panel factories for the Earth's deserts and the Moon, demonstration materials, Foetz, Luxembourg, 2023

22　ispace, Lunar Yard, Micro Rover, Luxembourg, Luxembourg, 2023
23　ispace, Lunar Yard, Luxembourg, Luxembourg, 2023

24   ispace, Lunar Yard, Luxembourg, Luxembourg, 2023

ON COSMIC ORDER

FRANCELLE CANE

The Moon is like a mountain. An object of scientific exploration for centuries, people have long sought to be the first to climb it. A summit that would have to wait a long time to be finally *conquered*. A floating, distant, empty rock—not all that fast. Like all other celestial bodies, the Moon is part of a cosmic order, a choreography that has been unfolding for billions of years. When the Moon, along with the Earth, joined in the dance, it had already been in full swing for a very long time.[1] Yet, as the James Webb Space Telescope delivers images of what a fragment of infinity looks like in eternity—reminding us, humans, of our dusty condition, fraught with fragility in a universe constituted of hundreds of billions of galaxies—the will to be the omniscient and omnipotent actors of the cosmos paradoxically persists. Today, more than ever, humans are developing a heightened "sense of spatial and temporal orientation," thanks to the support of "new technologies of surveillance, tracking, and targeting."[2] One of the most obvious phenomena is undoubtedly the "growing importance of aerial views: overviews, Google Map views, satellite views," which has led to "increasingly accustomed to what used to be called a God's-eye view," as filmmaker and author Hito Steyerl outlined a decade ago.[3] And if that view is the one we now have of any surface we can survey, it is no longer a question of knowing who will be the first to set foot on it—this has already happened and has since *made history*—but about close scrutiny and anticipation of who will be first to exploit it.

Although it does not belong to anyone in particular but to all of humanity, many people have claimed the Moon (or other celestial bodies such as the Sun) as their property—some have even sold plots of lunar land.[4] The dream is not new, and it is linked to the basic axiom of land ownership, laid down in the principle of property law: "Whoever's is the soil, it is theirs all the way to heaven and all the way to hell," Tim Ingold explains: "It means that, at least in theory, if I own a piece of ground, it is mine not just at the surface but upwards into the air, and downwards into the earth, as far as one can go."[5] A convention that, since the inception of property, has allowed landowners to make a profit from the sale of their deeds of ownership, earthly or cosmic. And if one wants to grasp, at least partially, some of the dynamics behind the logic of extractive exploitation both on Earth and on the Moon, an examination of this law is one of the key entry points.

This chapter aims at providing an introduction to the legal status of the Moon and celestial bodies, as well as the use of natural resources contained therein. It will address the legal framework applicable to and governing activities on celestial bodies, while paying particular attention to the study of the rules applicable to space resource utilization, specifically raising geopolitical and economic implications by highlighting legal parallels between earthly and non-earthly contexts that together form a *cosmic order*. As Bruno Latour and Peter Weibel attested, not without irony: "Shake the cosmic order and the order of politics will be shaken as well" — it is time to bring the cosmos back into political affairs.[6]

INTERNATIONAL SPACE LAW

In order to understand the rules that govern activities in space, it is necessary to introduce the legal context from which these rules are drawn — namely International Space Law. The latter is a part of Public International Law, a combination of principles and treaties establishing various sets of legal instruments and relations that are shared by members of the international community. International Space Law is substantially based on five separate United Nations treaties. A particular focus will be placed on the 1967 "Treaty on the Principles Governing the Activities of States in the Exploration and Use of Outer Space, Including the Moon and Other Celestial Bodies" (commonly referred to as the "Outer Space Treaty") and the 1979 "Moon Agreement." These two treaties establish the legal status of the Moon and other celestial bodies as well as the natural resources they contain.[7]

The Outer Space Treaty is notable within Public International Law, having been signed by 112 states and ratified by 89.[8] When it was launched, the Treaty was translated and published in the five official languages of the UN of the time: Chinese, French, English, Russian, and Spanish — Arabic was added when the language became the sixth UN official language in 1973 (FIG. 1). The process leading to the Outer Space Treaty began with a 1962 resolution that established some fundamental principles for regulating outer space activities. The Treaty was signed in January 1967 and came into force in October of that year, marking a decisive moment in International Space Law.[9] Indeed, as space law researcher Fabio Tronchetti put it, the Outer Space Treaty is the "cornerstone of International Space Law," since it constitutes the principal legal basis on which space activities have developed over the past six decades.[10]

The two decades following the Second World War were marked by efforts on both sides of the Iron Curtain to pursue the development of weapon-related technologies — particularly nuclear arms and rocket technology. The Outer Space Treaty was developed in a climate of diplomatic tension, as the US and Soviet

FIG. 1

Union were locked in a technological race over space capabilities. It was Sputnik I, the first artificial Earth satellite launched by the Soviets in 1957, that triggered a period of international debate about the development of a legal framework to regulate human activities in outer space.[11]

The following years were marked by a parallel process of decolonization around the world, allowing newly independent states to take part in international debates. Their presence pushed for a re-definition of the economic relations between the so-called global "North" and "South," and the declaration of a *New International Economic Order*, rightly stating that "the industrialized countries had taken advantage of the less-developed ones for centuries; therefore, it was time for the former to give something back to enable the latter to develop."[12] This had an impact on other international negotiations of the time, such as the UN Conventions of the Law of the Sea (UNCLOS) between 1974-1982. But by the beginning of the 1990s, the developing countries had (mostly) abandoned the idea of creating an international economic order that would entail "asymmetric obligations for the industrialized states" and instead "accepted a free-market approach to manage international areas."[13]

THE OUTER SPACE TREATY AND OTHER AREAS OF INTERNATIONAL LAW

What is particularly interesting on closer examination is that the economic and diplomatic context had a real impact on a set of agreements developed in the two decades following the drafting of the Outer Space Treaty. In fact, space law has a distinct relationship with environmental law, and with the legal frameworks of human rights.

One of the most essential notions of the Outer Space Treaty is probably found in the first sentence of its first article which attests that the Moon "shall be the province of all mankind" (FIG. 2). A phrasing that has been the cause of much debate to this day, with greatly varying interpretations of the status of the resources contained on and under the lunar surface. Just as there is no neutral material, there is no barren soil—especially when this soil is to be drilled.

While there is still no international treaty specifically governing the exploration and exploitation of the deep sea (a final draft was adopted by delegates of the Intergovernmental Conference on Marine Biodiversity of Areas Beyond National Jurisdiction [BBMJ] in March 2023), the UN Conventions on the Law of the Sea (UNCLOS) serve as a relevant example of a legal framework that incorporates the principles

ARTICLE I

The exploration and use of outer space, including the Moon and other celestial bodies, shall be carried out for the benefit and in the interests of all countries, irrespective of their degree of economic or scientific development, and shall be the province of all mankind.

Outer space, including the Moon and other celestial bodies, shall be free for exploration and use by all States without discrimination of any kind, on a basis of equality and in accordance with international law, and there shall be free access to all areas of celestial bodies.

There shall be freedom of scientific investigation in outer space, including the Moon and other celestial bodies, and States shall facilitate and encourage international cooperation in such investigation.

FIG. 2

and values enshrined in the Outer Space Treaty.[14] As such, under one of the articles (Article 136, Part XI), the deep seabed is declared to be the "common heritage of humankind," thus establishing "The Area" (the seabed beyond the national continental shelf) and governing the exploration and exploitation of resources of the deep seabed.[15] Like the Outer Space Treaty, the UNCLOS establishes a principle of non-appropriation under which no state can claim sovereignty over the deep sea or its resources. It also requires that resources in the deep sea be used for the benefit of all nations and affirms the equitable sharing of benefits from deep seabed mining. Another case that might at first glance seem far-fetched is the undeniable influence that outer space discussions have on the field of human rights (and vice-versa). Professor of International Law Steven Freeland, and Professor of International Space Law Ram S. Jakhu explain how the two fundamental international legal regimes for human rights — the International Covenant on Civil and Political Rights (ICCPR) and the International Covenant on Economic, Social and Cultural Rights (ICESCR) — were "finalized by the United Nations General Assembly and opened for signature on 16 December 1966, just a matter of a few weeks before the Outer Space Treaty (27 January 1967)." They point out that the first paragraph of the latter "demands that the exploration and use of outer space be 'for the benefit and in the interests of all countries, irrespective of their degree of economic or scientific development'," finally adding that "the fundamental concepts of space law are invariably linked with the fundamental concepts of rights and freedoms that we all (in theory) enjoy. In this regard, accepted principles of human rights law are highly relevant for our activities in space."[16] One may therefore wonder whether, by sharing the same struggles, the Latourian *earthly being* on one side and the *lunar being* on the other side are not ultimately the same *cosmic being*.

THE PRINCIPLE OF NON-APPROPRIATION

Another fundamental statement that underlines a key concept of space law is the second article of the Outer Space Treaty: in establishing the principle of non-appropriation, it emphasizes that outer space, including the Moon, cannot be owned or claimed by any country or entity (FIG. 3). There are two contradicting concepts in Roman law: *res nullius* and *res communes omnium*. While the former is used to describe a thing without an owner, the latter refers to a thing that is available to all and, therefore, can't be owned by anyone nor by a state. Thus, two schools of thought are established. The consequences of *res nullius* — where that which doesn't belong to anybody may be appropriated by anyone — were evident in Europe's imperial and colonial projects in Africa. Fortunately, it is the second principle, the *res communes omnium*, that gained general acceptance, and for good reason: "if States had been free to gain property rights on outer space by occupation or by other means, the risk of war between them would have increased significantly."[17]

In reality, the principles of non-appropriation and equitable benefit-sharing, which are key components of the Outer Space Treaty (as well as the Law of the Sea) have not been easy to enforce in reality: on Earth (and maybe soon also in outer space), powerful states and corporations have all too often been able to exploit natural resources in a way that benefits themselves at the tremendous expense of less powerful nations and communities, as well as the environment.[18] Aided by political and economic power dynamics, these entities have influenced the

ARTICLE II

Outer space, including the Moon and other celestial bodies, is not subject to national appropriation by claim of sovereignty, by means of use or occupation, or by any other means.

FIG. 3

development and interpretation of international agreements in their favor, resulting in a situation where the exploitative and inequitable use of natural resources has mostly benefited a small elite.

The role of the United Nations Committee on the Peaceful Uses of Outer Space (COPUOS), established in 1959, played a key role in defining the peaceful and legal status of outer space.[19] In particular, it was a decade after its creation that the committee's *raison d'être* became explicit: "After the successful Moon landing by the United States in 1969, awareness that Moon rocks might be returned to Earth and that mineral and other substances, as well as intangible resources, might be exploited, spread among the members of COPUOS."[20] And it is perhaps here that one comes to realize how some aspects of the Treaty did not anticipate the way certain terms, such as the

international permission under such a regime to actually start mining, but the comparison with the high seas for me means that you can certainly, in principle, extract mineral resources."[21] That's one small step for man, one giant leap for profit.

THE MOON AGREEMENT

Given the evolution of ever more precise and efficient technology, states decided to introduce a new legal instrument which would be commonly called the "Moon Agreement." Negotiated over seven years (1972-1979), the draft was eventually introduced by COPUOS in 1979 and entered into force after five more years, in 1984. Composed of 21 principles, the Moon Agreement reaffirms the non-appropriable nature of

> **ARTICLE III**
>
> States Parties to the Treaty shall carry on activities in the exploration and use of outer space, including the Moon and other celestial bodies, in accordance with international law, including the Charter of the United Nations, in the interest of maintaining international peace and security and promoting international cooperation and understanding.

FIG. 4

notion of *use*, would be defined, leaving an unsettled grey zone widely open to interpretation (FIG. 4). The concept of use has been exhaustively discussed and interpreted, especially in the context of resource exploitation and utilization. It represents a major issue in the debate: man set foot on the Moon only two years after the Treaty was put in place and, presumably, one could not anticipate what was to come. And this is where the shoe pinches, as space law professor Frans von der Dunk argues: "Unfortunately, the current status of international law, notably the 1967 Outer Space Treaty (which was drafted when nobody seriously thought this could happen), leaves some room for doubt. (...) One issue with the Outer Space Treaty (...) was that it had only declared outer space to be non-appropriable by states, essentially like the high seas. Some draw the conclusion from this that you then need an international regime and

the Moon and other celestial bodies as outlined in the Outer Space Treaty (FIG. 5). But here is the hitch: space-faring powers such as Russia, China, and the US have not signed up to it.[22] Article 11 of the Moon Agreement is probably the most talked about. The first paragraph mentions (for the first time) that the Moon and its natural resources represent a "common heritage of mankind"—a step forward from the Outer Space Treaty. The third paragraph seems to confirm that there are no private nor public property rights for any kind of entity on the surface or in the subsurface of the Moon. Furthermore, it also introduces "practical" rules regarding future missions on its site. It is perhaps its detailed features that do the most harm: "the intensive discussion of a possible revision of the Moon Agreement and in particular of article 11 is becoming topical, as numerous new realities of international and national space activities, including growing commercialization and the appearance of new actors, have had an impact on the interpretation of the concept of the 'common heritage of mankind'."[23] The controversy is therefore not specific to scientific exploration, but to the commercial exploitation of the natural resources of outer space. As far as science is concerned, there is a consensus on the right of states to extract and use extraterrestrial resources for research purposes (FIG. 6).[24]

Like colonized territories on Earth, lunar soil seems to be taken for granted, and is eventually seen as an "object of law."[25] Starting from a colonial logic, selenography appeared in the 16th century with the aim of studying the surface of the Moon in the same way that geography studies the Earth. By naming topographical features according to traditional geographical conventions, earthly nomenclatures were firmly established in this early race to the Moon,

ARTICLE 11

1. The Moon and its natural resources are the common heritage of mankind, which finds its expression in the provisions of this Agreement, in particular in paragraph 5 of this article.
2. The Moon is not subject to national appropriation by any claim of sovereignty, by means of use or occupation, or by any other means.
3. Neither the surface nor the subsurface of the Moon, nor any part thereof or natural resources in place, shall become property of any State, international intergovernmental or non-governmental organization, national organization or non-governmental entity or of any natural person. The placement of personnel, space vehicles, equipment, facilities, stations and installations on or below the surface of the Moon, including structures connected with its surface or subsurface, shall not create a right of ownership over the surface or the subsurface of the Moon or any areas thereof. The foregoing provisions are without prejudice to the international regime referred to in paragraph 5 of this article.
4. States Parties have the right to exploration and use of the Moon without discrimination of any kind, on the basis of equality and in accordance with international law and the terms of this Agreement.
5. States Parties to this Agreement hereby undertake to establish an international regime, including appropriate procedures, to govern the exploitation of the natural resources of the Moon as such exploitation is about to become feasible. This provision shall be implemented in accordance with article 18 of this Agreement.

FIG. 5

THE MOON'S CRITICAL ZONE

The products of the Cold War and global competition for technological and scientific supremacy raise questions in today's more complex and fragmented international environment, where issues of economic inequality and political polarisation appear to create significant obstacles to international regulation.

evoking seas, lands, lakes, and mountains through names such as *Mare Nectaris*, *Lacus Autumni*, *Montes Jura*, etc. Significantly, the development and production of selenographic and geological data today represents, more than ever, an assertion of geopolitical domination in what has become "an extended battleground for ongoing struggles over resource production and consumption."[26]

In her powerful 2018 essay, geographer Julie Michelle Klinger develops the implications of extractive activity on a global scale with great clarity. About a fundamental dimension of natural resource extraction on the Moon, she says: "this is what lunar mining advocates in the public and private sector desire: to transform the Moon into Earth's 'eighth continent'."[27] The last mountain to be climbed or the last frontier to be reached, the Moon represents one more territory to be expropriated. The as yet unresolved human challenge is to face and accept *ecological finiteness*—to appreciate a finite and non-extendable realm.[28] This calls for a new, non-earthly critical zone to be defined, a new subject to the cosmic order.

As cosmic beings, it seems credible that there should be a need for a new international cosmic regime. Beyond a global or planetary commons, it is the cosmic commons that must be urgently fought for—a new conception of our earthly and non-earthly political and legal relationship to natural resources. Although it seems utopian today to envisage global cooperation in earthly politics based on a form of economic restraint, (international) laws and institutions are the places of action where it will certainly be necessary to engage devices that can be levers to change our dependencies.[29] And let the mountain remain.

The mountain doesn't feel conquered or domesticated, contacted by civilization or incorporated into the human fold. It promptly forgets—if it ever noticed—that someone was up there, waving his arms ecstatically on the summit.[30]

ARTICLE 6

1 There shall be freedom of scientific investigation on the Moon by all States Parties without discrimination of any kind, on the basis of equality and in accordance with international law.
2 In carrying out scientific investigations and in furtherance of the provisions of this Agreement, the States Parties shall have the right to collect on and remove from the Moon samples of its mineral and other substances. Such samples shall remain at the disposal of those States Parties which caused them to be collected and may be used by them for scientific purposes. States Parties shall have regard to the desirability of making a portion of such samples available to other interested States Parties and the international scientific community for scientific investigation. States Parties may in the course of scientific investigations also use mineral and other substances of the Moon in quantities appropriate for the support of their missions.
3 States Parties agree on the desirability of exchanging scientific and other personnel on expeditions to or installations on the Moon to the greatest extent feasible and practicable.

FIG. 6

ENDNOTES

1. To date, most scientists agree that the universe was created and evolved about 13.7 billion years ago in an event called the 'Big Bang,' a state of high temperature and density. The Earth and Moon may have formed about 4.5 billion years ago, with the Moon forming a little later than the Earth. https://solarsystem.nasa.gov/
2. Hito Steyerl, "In Free Fall: A Thought Experiment on Vertical Perspective," *e-flux* Issue #24, April 2011, https://www.e-flux.com/journal/24/67860in-free-fall-a-thought-experiment-on-vertical-perspective.
3. Steyerl, "In Free Fall: A Thought Experiment on Vertical Perspective."
4. "Who owns the Moon?" Royal Museums Greenwich, https://www.rmg.co.uk/stories/topics/who-owns-moon.
5. Tim Ingold, *Correspondences* (Cambridge, UK: Polity Press, 2021): 94.
6. Bruno Latour, Peter Weibel, *Critical Zones: The Science and Politics of Landing on Earth* (Cambridge, MA: MIT Press, 2020): 13.
7. United Nations Office for Outer Space Affairs (UNOOSA), *International Space Law: United Nations Instruments* (New York: United Nations, 2017). The other three UN treaties include the 1968 "Rescue Agreement;" the 1972 "Liability Convention;" and the 1975 "Registration Convention".
8. United Nations Office for Disarmament Affairs (UNODA), "Treaty on the Principles Governing the Activities of States in the Exploration and Use of Outer Space, Including the Moon and Other Celestial Bodies," https://treaties.unoda.org/t/outer_space.
9. UNODA, "Treaty on the Principles Governing the Activities of States in the Exploration and Use of Outer Space, Including the Moon and Other Celestial Bodies."
10. Fabio Tronchetti, "Legal aspects of space resource utilization," in *Handbook of Space Law*, eds. Frans von der Dunk, Fabio Tronchetti (Cheltenham: Edward Elgar, 2015): 778.
11. Roger D. Launius, "Sputnik and the Origins of the Space Age," NASA, https://history.nasa.gov/sputnik/sputorig.html.
12. Steven Freeland, Ram S. Jakhu, "The intersection between space law and international human rights law", in *Routledge Handbook of Space Law*, eds. Ram S. Jakhu, Paul Stephen Dempsey (London: Routledge, 2017): 225.
13. Tronchetti, "Legal aspects of space resource utilization," 786.
14. "UN delegates reach historic agreement on protecting marine biodiversity in international waters," *UN News*, March 5, 2023, https://news.un.org/en/story/2023/03/1134157.
15. "Framework under international law: Part XI of the Convention," Umwelt Bundesamt, May 19, 2022, https://www.umweltbundesamt.de/en/topics/water/seas/deep-sea-mining/framework-under-international-law-part-xi-of-the.
16. Freeland and Jakhu, "The intersection between space law and international human rights law," 225–229.
17. Fabio Tronchetti, *The Exploitation of Natural Resources of the Moon and Other Celestial Bodies: A Proposal for a Legal Regime* (Leiden: Brill/Martinus Nijhoff Publishers, 2009): 11–15.
18. Karolaine Fainu, "'Shark calling': locals claim ancient custom threatened by seabed mining," *The Guardian*, September 30, 2021, https://www.theguardian.com/world/2021/sep/30/sharks-hiding-locals-claim-deep-sea-mining-off-papua-new-guinea-has-stirred-up-trouble.
19. United Nations Office for Outer Space Affairs (UNOOSA), Committee on the Peaceful Uses of Outer Space, https://www.uncosa.org/oosa/en/ourwork/copuos/index.html
20. Tronchetti, *The Exploitation of Natural Resources of the Moon and Other Celestial Bodies*, 38.
21. "Space Lawyers Are A Thing, And We Talked To One About The Future Of Cosmic Mining," Lila Shapiro in conversation with Frans von der Dunk, *The Huffington Post*, November 19, 2015, https://www.huffpost.com/entry/space-lawyers_n_564df48ae4b08c74b7349a05.
22. United Nations Office for Outer Space Affairs (UNOOSA), "Agreement Governing the Activities of States on the Moon and Other Celestial Bodies," https://www.unoosa.org/oosa/en/ourwork/spacelaw/treaties/intromoon-agreement.html.
23. Antonella Bini, "The Moon Agreement: Its effectiveness in the 21st century," *ESPI Perspectives* 14, October 2008, 5.
24. Michael J. Listner, "The Ownership and Exploitation of Outer Space: A Look at Foundation Law and Future Challenges to Current Claims," *Regent Journal of International Law 1*, 2003, 75.
25. Sarah Vanuxem, "Freedom through Easements," in *Critical Zones: The Science and Politics of Landing on Earth*, eds. Bruno Latour, Peter Weibel (Cambridge, MA: MIT Press, 2020): 240.
26. Julie Michelle Klinger, *Rare Earth Frontiers: From Terrestrial Subsoils to Lunar Landscapes* (Ithaca, NY: Cornell University Press, 2018): 201
27. Klinger, *Rare Earth Frontiers*, 230.
28. Gilles Clément, *Jardins, paysage et génie naturel* (Paris: Fayard, 2012): 31.
29. Pierre Charbonnier, *Abondance et Liberté: Une histoire environnementale des idées politiques* (Paris: La Découverte, 2020).
30. Ingold, *Correspondences*, 63.

ILLUSTRATIONS (FIG. 1–6)

1. "Outer Space Treaty" written in the six United Nations official languages.
2. Article I, *Treaty on the Principles Governing the Activities of States in the Exploration and Use of Outer Space, Including the Moon and Other Celestial Bodies*, United Nations, 1967.
3. Article II, *Treaty on the Principles Governing the Activities of States in the Exploration and Use of Outer Space, Including the Moon and Other Celestial Bodies*, United Nations, 1967.
4. Article III, *Treaty on the Principles Governing the Activities of States in the Exploration and Use of Outer Space, Including the Moon and Other Celestial Bodies*, United Nations, 1967.
5. Article 11, *Agreement Governing the Activities of States on the Moon and Other Celestial Bodies*, United Nations, 1979.
6. Article 6, *Agreement Governing the Activities of States on the Moon and Other Celestial Bodies*, United Nations, 1979.

Cameras,
tripods, lamps, and other studio equipment
seem to be an important part of the performance
of the mining technology itself.

Movements across the *empty* landscape
confuse the sight,

leaving space to wonder about the boundaries
between mediated and physical presence,
reality and simulation.

        The tracking and documentation
        become obsessive,
        going far beyond
        the technical image,
        projecting, instead,

        visions and missions of those who declare
        this space as their site.

Multiple,
but never conflicting, ways of seeing
contribute to this mechanism of laying claim.

Geologists claim the rocks as resources,
rovers claim the ground beneath the wheels,

engineers claim technology as the solution,
politicians claim new resource frontiers,

economists claim new markets
and financial opportunities, and
the Capitalocene claims the Moon.

The urgent need for a disclaimer.

Sometimes these claims hinge
on political ambitions,
sometimes on financial reveries,
and sometimes on the promises
of technological solutionism.

But they always operate
in conjunction with each other,
legitimizing and normalizing
the project of appropriation.

Its ultimate support system
is that of scientific investigation.

There is no discussion when you own the statistics.

The *facts* of science cut off the debate.

The time is running in a disorienting way.

It obscures the cycles
of day and night,
temporal patterns
of daily rhythms,
and even duration itself,
whether it is measured
in hours, centuries,
or billions of years.

Is it 21st July 1969?

Or just another nine-to-five
shift in a lunar laboratory?

Everything is organized around simulation.
Working for a hypothetical moment in the future,

    testing machines,
    testing humans,
    testing narratives,
    testing the limits of the law.

The demo becomes the ultimate *modus operandi*.

    This is a laboratory
    of space colonization dreams
    and the testing ground for
    technological,
    economic,
    and political tools
    to turn them into reality.

The bodies are grounded,
but imaginaries are floating.

Wrapped in the 21 layers of the A-7L space suit
made of beta cloth, or just a regular,
white laboratory coat;

> wheels coated with titanium wire mesh
> to prevent sinking into the soft soil,
> or dusty sneakers with fine basaltic sand
> finding its way into the smallest pore
> of the spectacle,

this is a display of the moving bodies
of the rovers and their masters.

The boundaries between the Moon and its earthly models are blurred. Designed to absorb the dreams of the Anthropocene, this lunar laboratory is both mise-en-scène and reality itself.

As theatre,

> it does not require
> the actual world
> it is narrating;
> it *is* the world
> in its own right.

But at the same time, as theatre,

> it is also prone to
> unexpected twists
> and turns, shifts
> of both the direction in which
> the story goes,
> and the narrators themselves.

THE EARTH'S MOON

FRANCELLE CANE AND MARIJA MARIĆ

From the development of human settlements on the Moon to asteroid mining of rare minerals and metals—the wild imaginaries of extraction-driven growth have, quite literally, transcended the boundaries of planet Earth. Next to the ambitious promises of infinite resources, techno-utopian visions of futuristic space settlements, or ludicrous geoengineering solutions to the climate crisis, such as the latest "moonshot" idea that proposes creating a solar shield around Earth by "ballistically ejecting" millions of tons of mined lunar dust—space, and the Moon in particular, has turned into the latest quick fix for earthly problems.[1] A triumph of technological solutionism, space mining has not only become the latest hot trend in the industrial discourse of sustainability and resource management, but also an important domain of financial investment and speculation, as well as political rhetoric. In this context, the Moon has been cast as the new resource frontier, a central stage for the projection of imagined profits, for nations and private companies alike.

The framing of the Moon as a new resource frontier has gained considerable traction during the past decade, as off-Earth mining advocates have used the so-called "rare earth crisis" to justify the establishment of a new, outer-planetary mining industry.[2] Departing from the already Anthropogenic appropriation of the Moon and other celestial bodies as the "province of all mankind" in the so-called *Outer Space Treaty*—a document on the exploration and use of outer space, adopted by the United Nations in 1967 and signed by more than one hundred nation-states so far—the current discourse surrounding space mining has, instead, focused on the private companies' visions of rare mineral finds, and the associated financial prospects.[3] Here, several countries, including the USA, Luxembourg, Japan, and the UAE, have used the ambiguities of the Outer Space Treaty's legal definitions, such as the one that prohibits ownership of lunar lands by any particular nation but does not prevent the exploitation of their soils (for the purposes of scientific exploration), effectively providing legal frameworks for private companies to freely mine celestial bodies, including the Moon. From NASA's recently announced call to buy up samples of lunar soil from private companies at a high price, recreating the effects of a gold rush, to Luxembourg re-branding itself (from a coal mining and steel mill economy) as an attractive destination for space mining start-ups, the frontrunners of the (neo-)liberalized take on extraterrestrial extraction have successfully positioned themselves as fertile ecosystems for the development of this new economy, leaving others to play catch-up with the fast-growing trend.[4]

This shift from nation-states to private industries as the leading stakeholders in the project of space mining has not, however, detached the project from nationalist rhetoric. The neocolonial and imperial outer space paradigm has found its place as a productive tool in the narratives of right-wing politicians, most notably Trump's 2016 "Make America Great Again" presidential campaign. As architectural historian Felicity D. Scott observes, the campaign sought to reverse the decision taken by NASA after the end of the Apollo missions in the early 1970s to focus its research on Earth rather than outer space, and instead re-position the US as the leader in space exploration—this time in a combination with openly nationalist rhetoric and with attention to the role of private companies such as Elon Musk's Space X and Jeff Bezos' Blue Origin.[5] In this constellation, as Scott further notes, the narratives of space exploration as a project of all of humanity and a tool for environmental protection gave way to narratives of crisis and impending wars, effectively legitimizing the central role of the military-industrial complex at the core of the entire enterprise.

Wrapped in promises of sustainability and the preservation of Earth by displacing the environmental destruction elsewhere—the greenwashing strategies of the nascent space mining industry have built their success as "solutions" to the crises of scarce resources and energy. With numerous reports and geological surveys outlining its abundance of rare metals and minerals, like the sought-after lithium that is required for the production of batteries for electric vehicles and other devices, the Moon is not only presented as a mine but also as a power plant. With energy costs being the biggest financial obstacle to the commercialization of Moon mining, the industry's focus on in-situ energy sourcing for both rovers and infrastructure to operate on lunar soil, as well as Earth-Moon transport and travel, and the detection of resources such is Helium-3 and water on the Moon, was greeted with great enthusiasm.[6] Even beyond its mineral wealth, the Moon's energy capacity was soon

1 Soviet technician working on the Sputnik 1 spacecraft, 1957

2   Apollo 17, View during the deployment of the transmitter antenna of the Surface Electrical Properties experiment, with Lunar Module Challenger, Schmitt, and the Lunar Roving Vehicle seen in the background

3   Signing of *Treaty on Outer Space*, January 27th, 1967

hailed as "the real discovery," as it allowed the visionaries of capitalist extraction to push the limits of the new frontier even further, all the way to Mars—repositioning the Moon, not as a destination, but rather as a technical, infrastructural energy hub or, in other words, a gas station.

The shifting of mineral exploitation from an exhausted Earth to its "invisible" hinterland—the Moon, celestial bodies, and finally, other planets—opens up a question: how does this new iteration of the space race, grounded in the promises of infinite resources, depart from the existing extractivist logic of capitalism and its destructive environmental and social effects on the ground? What are the materialities of space mining—its logistics, infrastructures, and workers—and what are their relationships to the existing geopolitical power hierarchies on Earth? How will the ongoing privatization of space, characterized by a turn to private companies as main actors in the exploitation of raw materials in space, affect the current status of extraterrestrial bodies as a kind of celestial commons? And finally, could we re-learn to look at the Moon beyond the abundance of its resources, beyond its framing as an *Other* and *outside*, seeing it instead as part of a shared struggle against capitalist dispossession?

THE MOON AS A NEW RESOURCE FRONTIER

Although it was not until the 1960s that humans managed to physically reach the Moon, the ground for its framing as a new resource frontier was prepared as far back as the early seventeenth century when the first lunar maps were drawn up. This was the period when selenography, the science organized around the study of the physical features of the Moon, gained traction, and technological advances such as the invention of the telescope brought about a more precise understanding of the lunar landscape. More than three centuries of mapping, classification, and nomenclature development followed before the Apollo 11 Mission of 1969 enabled the first physical survey of the Moon's soils. The scientific rationalization of the Moon and its natural resources went hand-in-hand with the evolution of the capitalist economy, securing its place in the domain of the Anthropocene.

But in order to be framed as resource-rich, the Moon first needed to be labeled as a void. In her book *Friction: An Ethnography of Global Connection*, anthropologist Anna Tsing defines frontiers as "zones of not yet; not yet mapped, not yet regulated."[7] She argues that contrary to the narratives of their proponents, new resource frontiers are not discovered but rather constitute projects in the making, shaped by processes of translation, transformation, regulation, and control.[8] The "magical vision" behind the construction of frontier regionality, as she further points out, has the power to blur fiction and reality, asking the participants to see landscapes which do not yet exist, or unsee the existing ones, all while covering up the conditions of its own production.[9] Labeling the Moon as "empty" before detecting it as full (of metals and minerals) could thus be seen as a process that has a long history in colonial practices on Earth. However, as Danika Cooper reminds us in her study of the construction of deserts as "empty" landscapes, "emptiness is neither a geographical category nor an ecological feature, it is a culturally constructed, political instrument," while the very process of mapping regions as "empty" could be seen as a way of obliterating their historical and political densities while at the same time validating their social and ecological exploitation and dispossession.[10]

In the same way that extractive imaginaries of the capitalist economy keep the demarcation of frontiers fluid, always ready to expand and absorb what is around it, or travel yet further—the limits and the scale of the mine itself have become fluid under the conditions of capitalism. Building upon Mazen Labban's concept of the "planetary mine" not only as a site of extraction but also as an extractive condition that operates within the circulation of capital and extends across the entire geography of the Earth, Martín Arboleda offers a new reading of the mine as "not a discrete sociotechnical object but a dense network of territorial infrastructures and spatial technologies vastly dispersed across space."[11] This expanded understanding of the mine, in the context of global late capitalism, forces us to look at "extractive industries beyond the mere wrestling of minerals from the soil," and to recognize a broader set of intertwined processes and infrastructures, including technological development, financial speculation, and labor exploitation that contribute to the pertaining logic and practices of mining, whether on Earth or beyond it.[12]

4   Apollo 11 Data Acquisition Camera

5   Apollo 11, the first photo by Neil Armstrong after setting foot on Moon

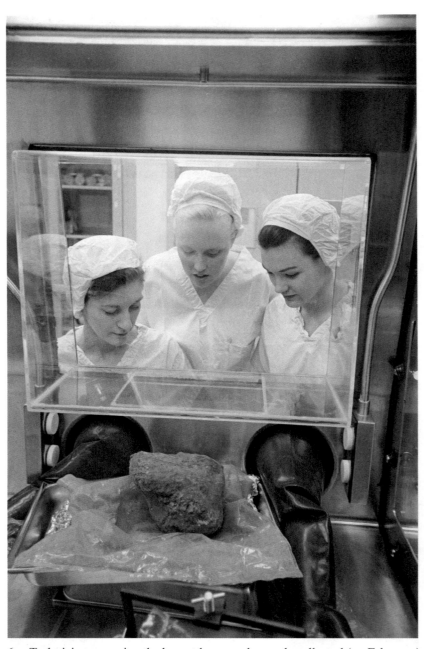
6   Technicians examine the largest lunar rock sample collected (24 Feb. 1971)

7   Astronaut Michael Collins (right) tours the Lunar Receiving Laboratory (LRL) in 1967, including the vacuum gloveboxes

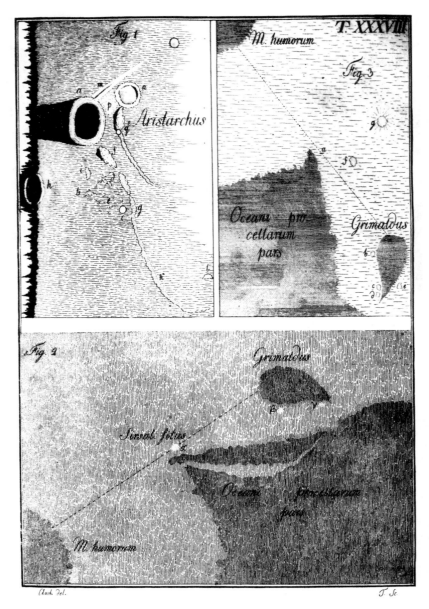

8 Johann Hieronymus Schröter, *Selenotopographische Fragmente* (extract), 1791–1802

In other words, if the mine cannot be reduced to the site of extraction, it is important to understand that the extracted materials themselves can never be understood as "raw," and detached from their social, economic, political, and environmental histories. Describing the science of geology as a "racialized optic razed on the Earth," Kathryn Yusoff suggests that the history of geological extraction should be read together with the history of slavery, arguing that, under the paradigm of "extractive geo-logic," both extract value from what is considered to be non-human: rocks and slaves.[13] Existing as a mode of both accumulation and dispossession, "depending on which side of the geologic color line you end up on," geological classification provides lenses to look at entire regions, communities, and finally, worlds, as elements in a system of monetary value.[14] Its visions of emptiness, declaring what is living and what is not, the construction of "no-thing," dehumanization of the colonized territories, however, goes beyond the boundaries of the Earth, setting the grounds for what might be described instead as an "extractive cosmo-logic."[15] By becoming the object of geological research, the Moon has been effectively colonized, and naturalized as a site of extraction.

BEYOND THE COMMONS:
THE MOON AS A COMRADE

This poses the question: what if we were to zoom out, and tried to see the Moon outside the lens of the Anthropocene? Can we start looking at the Moon beyond its constructed resource-richness, regardless of whether some or all have access to it, and instead take the position of what Oxana Timofeeva describes as "cosmic solidarity," referring to the "recognition that the Sun is neither a master, nor a slave," but rather "a comrade?"[16] To do that, we must not only challenge the capitalist imaginaries of growth and profit that legitimize ceaseless expansion, extraction, and destruction but go a step further, taking a leap *beyond* the commons—itself a form of use and governance of resources—and thereby setting limits to the Anthropocene.[17] In his 1972 article "Should Trees Have Standing? Towards Legal Rights for Natural Objects," Christopher D. Stone outlined principles for what would later become known in the legal sphere as "environmental personhood"—a legal concept organized around "designating parts of nature as legal persons entitled to independent regard and consideration."[18] Drawing on the established recognition of non-human entities, such as corporations, as bearers of legal rights, environmental personhood appeared to serve as a legal tool for limiting the human impact on the environment. It did, however, perpetuate another problematic position by attaching the notion of "personhood" to property relations. Observing how "until the rightless thing receives its rights, we cannot see it as anything but a thing for the use of 'us'—those who are holding rights at the time," Stone wrote: "If a human being shows signs of becoming senile and has affairs that he is de jure incompetent to manage (...) the guardian (or 'conservator' or 'committee,' the terminology varies) then represents the incompetent in his legal affairs."[19]

Already subsumed by the colonizing ideologies of global capitalist machinery and its military-industrial complexes (regardless of their geopolitical affiliations), the Moon needs more than a guardian. In rethinking the current narratives of space mining, we need to think beyond preservation and equal access to lunar resources, challenging not only the lens we use to look at the Moon but the very gaze itself. Here, one could ask what it would mean to see the Moon not as a new resource frontier, not even as the property of "all of humanity" or a form of global commons, but rather as a comrade—as one with whom we join in the struggle against capitalist extraction? Writing on the notion of *comradeship*, Jodi Dean observed: "The term 'comrade' points to a relation, a set of expectations for action. It doesn't name an identity; it highlights the sameness of those who share a politics (...) Comradeship isn't personal. It's political."[20] Originating from the Latin word *camera*, referring to a "room" or "chamber," the spatiality embodied by comradeship—that of sharing space, living under the same roof—reminds us that the Moon and the Earth move in the same orbit, both in galactic and political terms.[21] It forces us to rethink the form of our alliances, setting the ground for an alternative kind of political agency: that of cosmopolitics, necessary for repairing and decolonizing our earthly and beyond-earthly imaginaries.[22]

9   Lunar Sample 61016, known as 'Big Muley'; named after Bill Muehlberger, the leader of the Apollo 16 field geology team

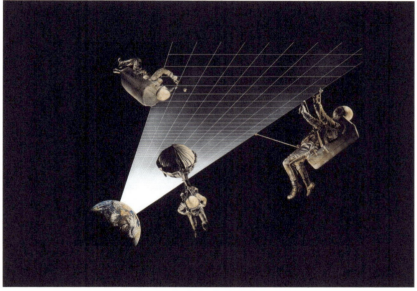

10 Alessandro Poli (Superstudio). *Autostrada Terra-Luna [Earth-moon highway]*, 1970–1971. Alessandro Poli fonds, Collection Canadian Centre for Architecture, Montréal

Cette plaine ne serait qu'un immense ossuaire. (Page 93.)

11 Jules Verne, *Autour de la lune* (extract p.88); 44 drawings by Emile Bayard and A. de Neuville; engraved by Hildibrand, 1893, 1 vol.

IMAGE CREDITS

1   Copyright: Sovfoto/Universal Images Group/via Getty Images
2   Source: NASA/JSC/USRA/Lunar and Planetary Institute
3   Copyright: UN Photo
4   Copyright: Image by Dane Penland, National Air and Space Museum, Smithsonian Institution
5   Source: NASA
6   Source: NASA
7   Source: NASA
8   Copyright: Bibliothèque de l'Observatoire de Paris
9   Source: NASA/JSC/USRA/Lunar and Planetary Institute
10   Copyright: Archivio Superstudio
11   Source: Bibliothèque nationale de France, département Littérature et art

ENDNOTES

1. Oliver Milman, "A Solution to the Climate Crisis: Mining the Moon, Researchers Say," *The Guardian*, February 8, 2023, https://www.theguardian.com/science/2023/feb/08/moon-dust-moonshot-geoengineering-climate-crisis.
2. See Julie Michelle Klinger, *Rare Earth Frontiers: From Terrestrial Subsoils to Lunar Landscapes* (Ithaca: Cornell University Press, 2018).
3. United Nations Office for Outer Space Affairs, "Treaty on Principles Governing the Activities of States in the Exploration and Use of Outer Space, Including the Moon and Other Celestial Bodies," 1967, Article 1.
4. Following the global steel crisis of the 1970s and 1980s', Luxembourg's post-industrial national strategy focused on the knowledge and service economy, as well as the investment into technology. Already in 1985, the country positioned itself in the space business, by launching the Société Européenne des Satellites (SES), currently the satellite company with the largest revenue in the world. See *Space Mining Is Here, Led by This Tiny Country* (Bloomberg, 2021), https://www.youtube.com/watch?v=uC012myvQs&t=169s&ab_channel=BloombergOriginals and Edward Helmore, "Nasa Is Looking for Private Companies to Help Mine the Moon," *The Guardian*, September 11, 2020, https://www.theguardian.com/science/2020/sep/11/nasa-moon-mining-private-companies.
5. Felicity D. Scott, "Self-Government on the High Frontier," in *Futurity Report*, ed. Eric C. H. de Bruyn and Sven Lütticken. (Sternberg Press, 2019), 138–58: 140.
6. European Space Agency, "Helium-3 Mining on the Lunar Surface," Organization's Website, accessed February 26, 2023.
7. Anna Tsing, *Friction: An Ethnography of Global Connection* (Princeton and Oxford: Princeton University Press, 2005), 28.
8. Anna Tsing, *Friction: An Ethnography of Global Connection* (Princeton and Oxford: Princeton University Press, 2005). See also Claude Raffestin, "Elements for a Theory of the Frontier," trans. Jeanne Ferguson, *Diogenes 34*, no. 134 (June 1, 1986): 1–18.
9. Tsing, *Friction*.
10. Danika Cooper, "Drawing Deserts, Making Worlds," in *Deserts Are Not Empty*, ed. Samia Henni (New York: Columbia Books on Architecture and the City, 2022), 73–108: 83.
11. Martín Arboleda, *Planetary Mine: Territories of Extraction under Late Capitalism*, Kindle Edition (London: Verso, 2020) and Mazen Labban, "Deterritorializing Extraction: Bioaccumulation and the Planetary Mine," *Annals of the Association of American Geographers* 104, no. 3 (2014): 560–76.
12. Arboleda, *Planetary Mine*.
13. Kathryn Yusoff, *A Billion Black Anthropocenes or None* (Minneapolis: University of Minnesota Press, 2018).
14. Yusoff, *A Billion Black Anthropocenes or None*.
15. Oxana Timofeeva, *Solar Politics* (Cambridge: polity, 2022). See also Sophia Roosth, "Life, Not Itself: Inanimacy and the Limits of Biology," *Grey Room*, no. 57 (2014): 56–81.
16. Timofeeva, *Solar Politics*.
17. On the definition of commons see Yochai Benkler, "The Political Economy of Commons," *Upgrade* 4, no. 3 (2003) and Elinor Ostrom, *Governing the Commons: The Evolution of Institutions for Collective Action* (Cambridge University Press, 1990).
18. Gwendolyn J. Gordon, "Environmental Personhood," *Columbia Journal of Environmental Law* 43, no. 1 (2018): 49–91, 49.
19. Christopher D. Stone, "Should Trees Have Standing?—Towards Legal Rights for Natural Objects," *Southern California Law Review*, no. 45 (1972): 450–501, 464.
20. Jodi Dean, "Four Theses on the Comrade," *E-Flux Journal*, no. 86 (November 2017), https://www.e-flux.com/journal/86/160585/four-theses-on-the-comrade/.
21. On the etymology of the word 'comrade' see Jodi Dean, *Comrade: An Essay on Political Belonging*, Kindle Edition (London and New York: Verso, 2019).
22. See Rania Ghosn and El Hadi Jazairy, "Cosmorama: A Peep Show of the New Space Age," *New Geographies*, no. 11 (2020): 157–69. and Isabelle Stengers, "The Cosmopolitical Proposal," in *Making Things Public: Atmospheres of Democracy*, ed. Bruno Latour and Peter Weibel (Karlsruhe and Cambridge: ZKM and The MIT Press, 2005), 994–1004.